THE SOLAR SYSTEM

Please visit our web site at: www.garethstevens.com
For a free color catalog describing Gareth Stevens Publishing's list of high-quality books and multimedia programs, call 1-800-542-2595 (USA) or 1-800-387-3178 (Canada). Gareth Stevens Publishing's fax: (414) 332-3567.

Library of Congress Cataloging-in-Publication Data

Spin.
The solar system.
p. cm. — (Discovery Channel school science. Universes large and small)
Summary: Discusses our solar system, including its planets, moons, asteroids, comets, and more.
ISBN 0-8368-3372-4 (lib. bdg.)
1. Solar system—Juvenile literature. [1. Solar system.] I. Title. II. Series.
QB501.3.S67 2003
523.2—dc21 2003042499

This edition first published in 2004 by
Gareth Stevens Publishing
A World Almanac Education Group Company
330 West Olive Street, Suite 100
Milwaukee, WI 53212 USA

Further resources for students and educators available at www.discoveryschool.com

Designed by Bill SMITH STUDIO
Creative Director: Ron Leighton
Design: Sonia Gauba, Brian Kobberger, Jay Jaffe, Eric Hoffsten, Nick Stone
Photo Editor: Justine Price
Production Director: Peter Lindstrom
Art Buyer: Lillie Caporlingua
Print consulting by Debbie Honig, Active Concepts

Gareth Stevens Editor: Betsy Rasmussen
Gareth Stevens Art Director: Tammy Gruenewald
Technical Advisor: Russell Berg

Printed in the United States of America

1 2 3 4 5 6 7 8 9 07 06 05 04 03

Writer: Lorraine Hopping Egan

Editors: Jackie Ball, Bill Doyle, Marc Gave

Content Reviewer: Dr. Stephen Maran

Photographs: Cover, © Discovery/The Image Bank; p. 2, pp. 4-5, © Sally Bensusen/Science Source/Photo Researchers; p. 9, © Discovery/Corbis-Bettmann; p. 12, Mercury, Science Source/Photo Researchers; left, © John Sanford/Photo Network; right, U.S. Geological Survey (U.S.G.S.); p. 13, NASA; p. 18, NASA; p. 19, Pluto and Charon, NASA; Clyde Tombaugh, Science Photo Library/Photo Researchers, p. 21, all photos, NASA; p. 22, David Ducros/Science Source/Photo Researchers; p. 24, NASA/Science Photo Library; p. 26, Mark Twain, © Discovery Archive Photos; pp. 26-27, Halley's Comet, © Discovery/The Image Bank; p. 28, Carolyn Shoemaker, © David Parker/Science Source/Photo Researchers; Comet-Shoemaker-Levy 9, © MSSSO, ANU/Science Photo Library/Photo Researchers.

Illustrations: pp. 10-11, 14-15, Alexis Seabrook; p. 25, NASA map adapted by Joe Le Monnier.

Acknowledgements: Interview with Carolyn Shoemaker reproduced by permission. © 1993 Astronomy Magazine, Kalmbach Publishing Company.

Sit down. Strap in. And get set to blast off! Where are we going? The Solar System. Don't worry, we don't have far to go. Why? Because we are already there.

Our place in the Solar System is the most important part of our address—even though we never write it on an envelope. Earth is just one of nine planets that revolve around the Sun. And then there are lots of moons, asteroids, and other bodies, too.

Hop on board as Discovery Channel take you on a SOLAR SYSTEM mission. Space awaits!

THE SOLAR SYSTEM

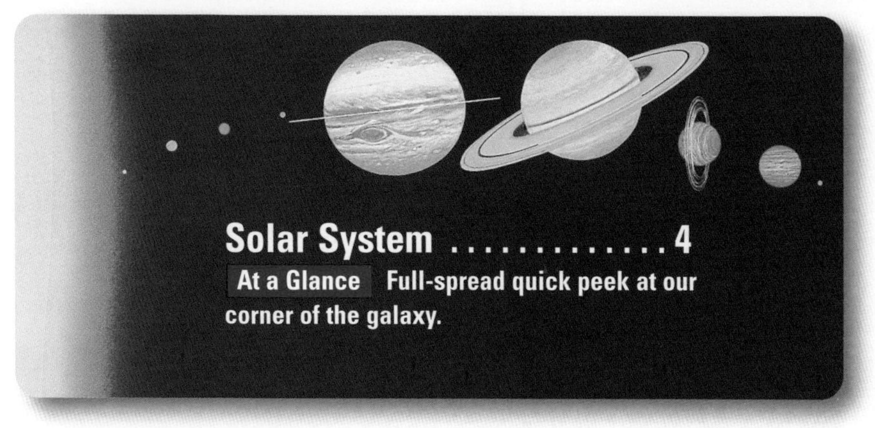

Duck!! See page 24.

Final Project

Solar System

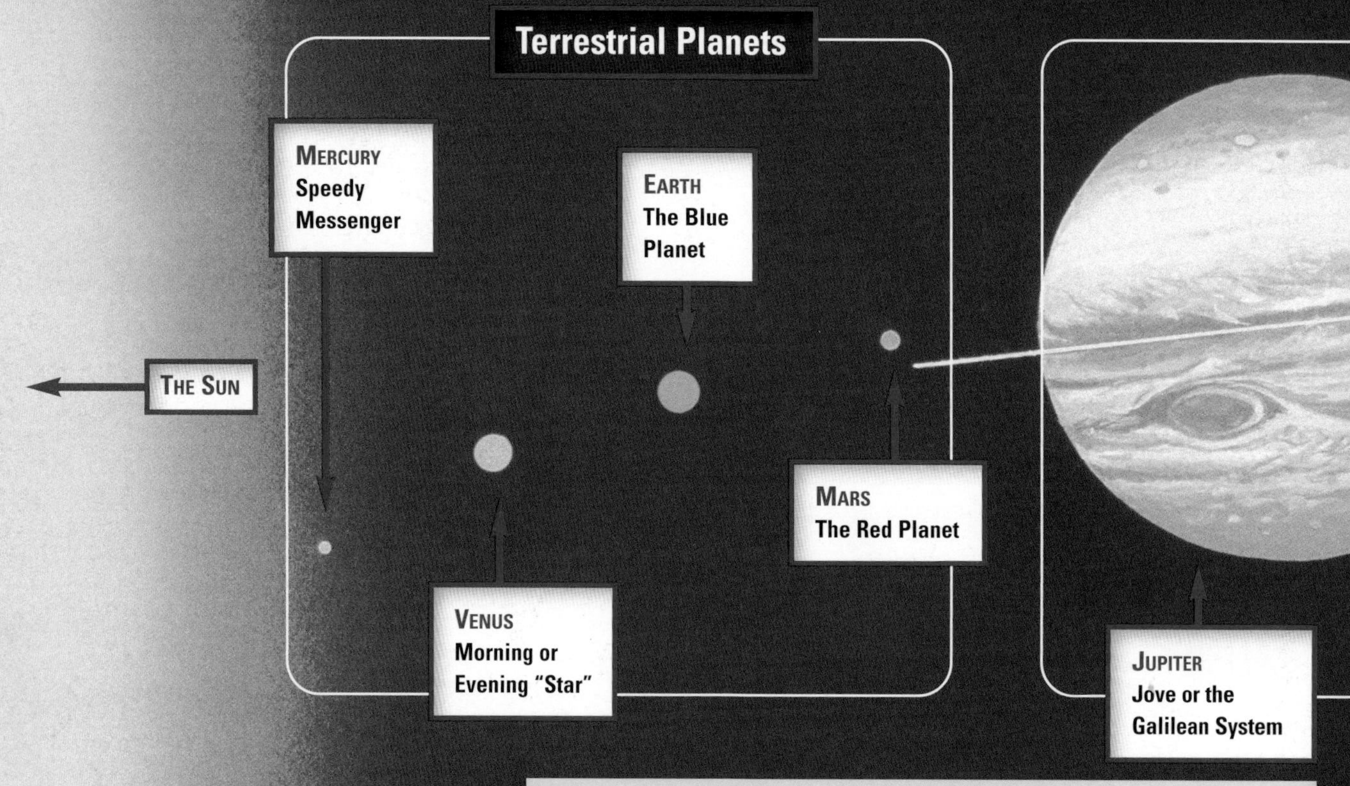

The true superstar of our Solar System isn't a planet—it's a star! The one and only star we have in our system. And we call that star the Sun. *Solar System* means "system of the Sun."

Like any celebrity, the Sun attracts other bodies with its strong gravitational pull. And our Sun is huge. If it were hollow, it could hold 1.3 million Earths.

Of course, our Solar System is more than just the Sun. It is the Sun and everything the Sun's gravity can grab: nine planets and their moons, asteroids, and comets. In addition to providing the gravitational glue that binds the Solar System, our massive Sun is the nuclear power plant that supplies it with energy. Without the Sun's energy, life on Earth could not exist.

The Sun makes up more than 99 percent of all the mass in the Solar System. That means all the other bodies together make up less than one percent. No wonder it's the center of attention.

Lost in Space?

Can't remember the nine planets and their order? Just say this sentence:

My Very Excellent Mom Just Served Us Nice Pickles.

The first letter of each word matches the first letter of each planet: Mercury, Venus, Earth, Mars, Jupiter, Saturn, Uranus, Neptune, and Pluto.

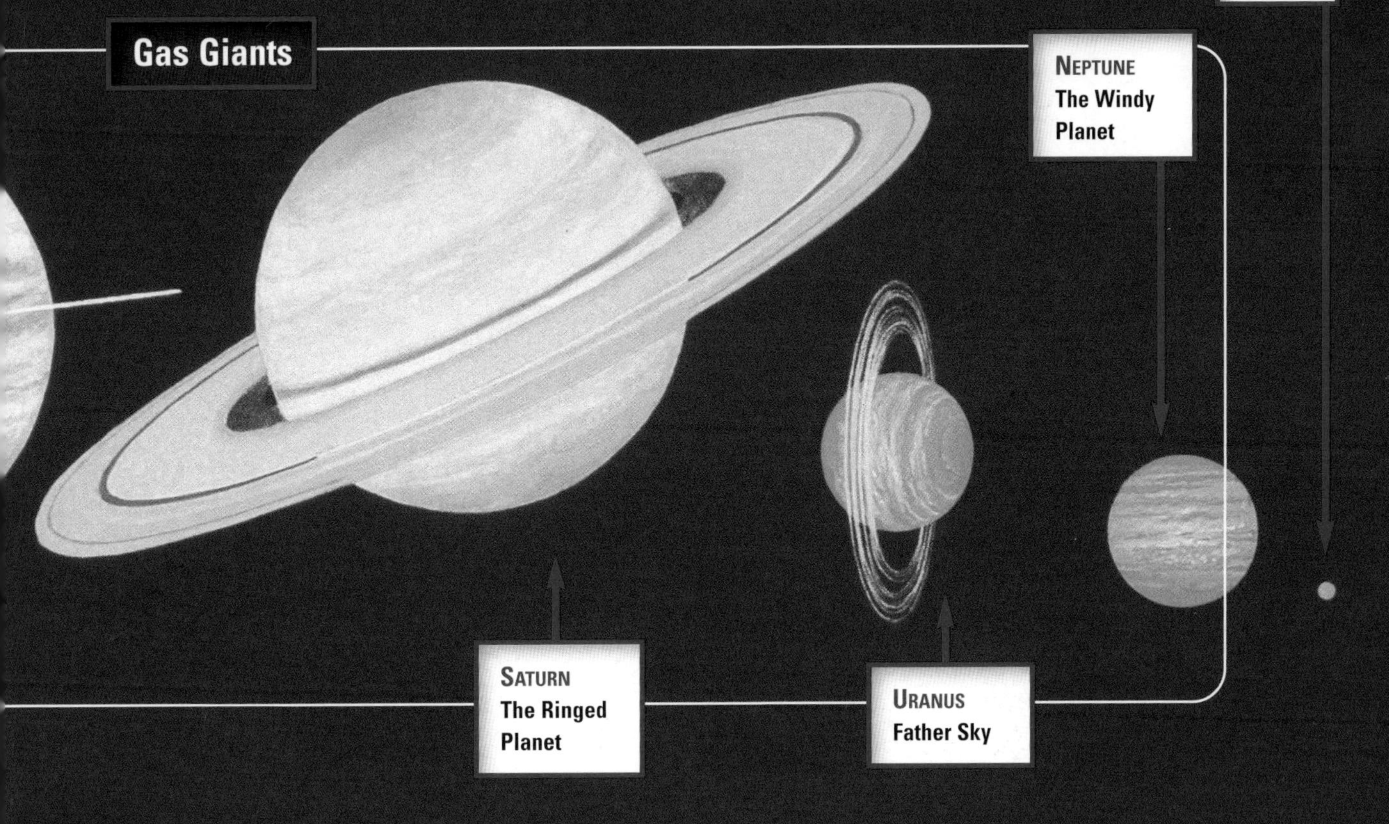

STAR—A huge body made of hot luminous gas that is held together by its own gravity. Compared to other stars, our Sun is a star of average size and energy.

PLANET—A large body that orbits a star. Terrestrial (land) planets have rocky cores and are much smaller than gas giants, which have thick, gassy atmospheres.

ASTEROID (OR MINOR PLANET)—A body smaller than a planet and made of rock, metal, or other minerals. There is a large belt of asteroids between Mars and Jupiter.

MOON (OR SATELLITE)—A body that orbits a planet or asteroid. Most moons are much smaller than their planet or asteroid.

COMET—An object composed of frozen gases and dust particles that was formed on the outskirts of the Solar System. Comets orbit the Sun, and most have long, skinny orbits that take them beyond the planets.

ELLIPSE—An elongated circle; the shape of a closed orbit; the path that one celestial body makes around another.

METEOR (OR SHOOTING STAR)—A chunk of rock or metal that enters Earth's atmosphere and burns up in a fiery show.

METEORITE—A chunk of rock or metal that has fallen to Earth from beyond its atmosphere.

METEOROID—A chunk of rock or metal moving through space.

GRAVITY—The force that keeps celestial bodies in place. The Sun's gravity keeps the planets in orbit.

Solar System

Time for a little planet-hopping!

Check out the chart below for facts on Earth and the other eight planets in our Solar System.

PLANETS	DIAMETER	YEAR (ONE ORBIT AROUND SUN) IN EARTH TIME	DAY (ONE ROTATION ON AXIS) IN EARTH TIME	MAIN GASES IN ATMOSPHERE
MERCURY	3,029 miles (4,874 km)	88 days	59 days	Traces of argon, neon, helium
VENUS	7,515 miles (12,091 km)	224 days, 17 hours	243 days	Carbon dioxide (very thick)
EARTH	7,921 miles (12,745 km)	365 days, 6 hours	1 day, 23 hours, 56 minutes	Nitrogen, oxygen
MARS	4,218 miles (6,787 km)	687 days, 23 hours	24 hours, 37 minutes	Carbon dioxide (thin)
JUPITER	88,793 miles (142,868 km)	11 years, 11 months	9 hours, 56 minutes	Hydrogen, helium
SATURN	74,853 miles (120,438 km)	29 years, 5 months	10 hours, 39 minutes	Hydrogen, helium
URANUS	31,744 miles (51,076 km)	84 years	17 hours, 14 minutes	Hydrogen, helium, methane
NEPTUNE	30,757 miles (49,488 km)	164 years	16 hours, 7 minutes	Hydrogen, helium, methane
PLUTO	1,430 miles (2,301 km)	248 years	6 days, 9 hours, 18 minutes	Possibly nitrogen, carbon dioxide, and/or methane

WHAT A TRIP!

Searching for that fun-in-the-Sun vacation spot? Check out the chart below. It shows the average distance of each planet from the Sun.

MERCURY	36 million miles (58 million km)	JUPITER	480 million miles (772 million km)
VENUS	67 million miles (108 million km)	SATURN	890 million miles (1,432 million km)
EARTH	93 million miles (150 million km)	URANUS	1,790 million miles (2,880 million km)
MARS	140 million miles (225 million km)	NEPTUNE	2,800 million miles (4,505 million km)
		PLUTO	3,680 million miles (5,920 million km)

Stats

Mathwise

HOW MUCH IS 1,790 MILLION?
It's the same as 1 billion, 790 million.

NUMBER OF MOONS	EARTH WEIGHT OF 110-POUND (50 KG) KID	CLAIM TO FAME
0	41.0 pounds (18.5 kg)	Daytime temperature 880°F (470°C); nighttime –300°F (–183°C).
0	97.0 pounds (44.0 kg)	Rotates east to west instead of west to east, as the other planets do.
1	110.0 pounds (50.0 kg)	Only body known to have life; mild climate; water as solid, liquid, gas.
2	42.0 pounds (19.0 kg)	Ice-capped poles; four seasons; global dust storms.
At least 16	276.5 pounds (125.5 kg)	Colorful bands of gas, marked by lightning, with auroras near poles; Great Red Spot is a storm bigger than Earth.
At least 18	118.0 pounds (53.5 kg)	Circled by colorful ice rings ranging in width from very narrow to house-size.
17	102.5 pounds (46.5 kg)	Tipped sideways so pole—not equator—faces Sun.
8	135.5 pounds (61.5 kg)	Wind speeds of 1,500 mph (2,414 kph)—highest recorded on any planet.
1	5.5 pounds (2.5 kg)	Tiny size; offbeat orbit; some people argue it's not even a planet.

Activity

SPACED OUT! Got an orange and access to a football field? Then you can create your own Solar System. Put the orange on one goal line. This is the Sun. Using the chart above and a scale of 1 yard = 10 million miles, Mercury is a speck of dust about halfway between the 3 and the 4 yard lines. Now locate the positions of the rest of the planets. (Although the yard lines on a football field go from 0 to 50 and back to 0, treat them as 100 continuous yard lines.)

Can you score a touchdown and place all the planets on the field? What yard line do you land on? Is there even enough room? If not, what scale would you have to use to fit all the planets on the same football field? (If you can't get to a football field, no problem. Use graph paper. One unit = 10 million miles. Use as many sheets as you need.)

Bonus: The farther the planets are from the Sun, the more spaced out they are. Figure out how many times farther each planet is from the Sun than the previous one.

Greece, 384–322 B.C.

During his lifetime, the Greek philosopher Aristotle thinks and thinks. Finally, he decides that Earth is the center of all existence and that the other heavenly bodies revolve around it in perfect circles. People are happy. We just love this idea of a geocentric (Earth-centered) universe. It makes us feel important.

Alexandria, Egypt, A.D. 127–145

Claudius Ptolemaeus, known as Ptolemy, refines Aristotle's ideas but agrees that everything revolves around Earth. People remain happy. After all, it's only right that everything should travel around us. We're that special. Besides, there are references in the Bible that seem to back up geocentricity.

Poland, 1543

Astronomer Nicolaus Copernicus develops a theory that Earth and the other planets orbit the Sun—that they are heliocentric (Sun-centered). What? We're not the center of attention? People are angry. Copernicus doesn't feel their wrath; he dies right after his ideas are published.

Padua, Italy, January 7, 1610

Has Galileo Gone Loony?

by Signore Scoop

It's a beautiful night. You look up at a bright round shiny Moon. A perfect sphere, right? Wrong. Or so said Galileo Galilei today in Padua.

Professor Galileo says our pure Moon is not perfect. He claims he has seen peaks and valleys on the Moon—and that they are bigger than Earth's peaks and valleys! That would mean the moon might be more special in some ways than Earth.

But wait, there's more! Galileo also says that four little stars circle Jupiter. Stars circling another planet? But everything is supposed to circle around us, the Earth! Has Galileo not learned the lessons of Ptolemy? The Church and all other scholars tell us all the stars orbit our noble planet. It's us, us, us! Everything is about us! After all, anyone can plainly see the Sun and stars cross our sky each day and each night!

But Galileo says he has proof in the form of an amazing instrument. It's called the telescope, and objects seen through it are magnified more than a thousand times. Planet Earth, we must now consider, might be more ordinary than extraordinary.

Let me speak first of the surface of the Moon. The brighter part and the darker part are plain to everyone, and every age has seen them. Other spots, smaller in size, sprinkle the whole surface of the Moon. These spots have never been observed by anyone before me.

From my observations, I feel sure that the surface of the moon is not perfectly smooth and exactly spherical, as a large school of philosophers considers. On the contrary, it is uneven, like the surface of the Earth itself. The boundary that divides the part in shadow from the enlightened part is marked by an irregular, uneven, and very wavy line.

There appear very many bright points within the darkened portion which gradually increase in size and brightness. After an hour or two, they become joined to the rest of the bright portion.

Now, is it not the case on the Earth before sunrise, that while the plain is still in shadow, the peaks are illuminated by the Sun's rays? Does not the light spread further and, when the Sun has risen, do not the illuminated parts of the plains and hills join together?

The grandeur of such prominences [peaks] and depressions [valleys] in the Moon seems to surpass the ruggedness of the Earth's surface.

Jupiter Has Company

On the 7th day of January in the present year, 1610, the planet Jupiter presented itself to my view. I noticed . . . three little stars, small but very bright, were near the planet. They seemed to be arranged exactly in a line and to be brighter than the rest of the stars. On the east side there were two stars, and a single one towards the west.

At first I believed them to be fixed stars. But on January 8th, I found a very different state of things. There were three little stars, all west of Jupiter, and nearer together than on the previous night.

I became afraid lest the planet might have moved differently from the calculation of astronomers. But on January 10th, there were two stars only, and both on the east side of Jupiter. The third, as I thought, [was] hidden by the planet.

I knew that changes of position could not by any means belong to Jupiter, but to the stars. I therefore concluded that there are three stars moving about Jupiter. [Further] observations established that there are not only three, but four.

Activity

SCOPE IT OUT Galileo's telescope was very advanced for its day, but it's primitive compared to what we have today. Circling in space is the Hubble Space Telescope. You can log onto its web site at hubble.nasa.gov for some incredible close-ups of the planets and stars.

A Sun-sational Journey

Galileo didn't stop with the Moon. He turned his telescope toward the Sun and spied dark blemishes called sunspots traveling across the face of the supposedly pure Sun. He used this discovery as part of his support for Copernicus's theory that the planets orbit the Sun.

Public outrage was incredible. Nearly everyone accepted the Church's teaching that the Sun, a perfect sphere, orbits Earth. Galileo was put on trial for publishing his "heretical" observations. Under threat of torture, he renounced his ideas and spent the rest of his life under house arrest.

Eventually, Galileo went blind, probably from looking directly at the Sun. We've come up with a safer way for you to explore the Sun. Just take our tour below and you'll be a Sun-seasoned traveler in no time.

1 THE ULTIMATE SARDINE CAN We start our journey in the very core of the Sun. And you? You're an atom of hydrogen, the smallest and simplest of all atoms in the universe. At this location, half of the Sun's mass is squeezed into a fraction of its volume. That means a tremendous amount of pressure. How does it feel? Imagine that you're a sardine in a can being stomped by an elephant carrying tons of bricks. Okay? Now multiply that feeling by a zillion.

The pressure and heat (15,000,000 °K!) fuse you and another hydrogen atom into an atom of helium. But you're not alone—hundreds of millions of tons of hydrogen are fusing into helium every second.

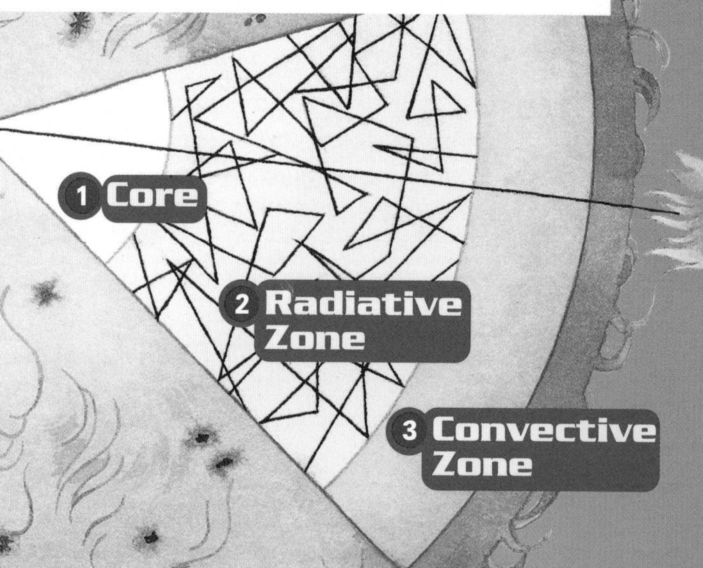

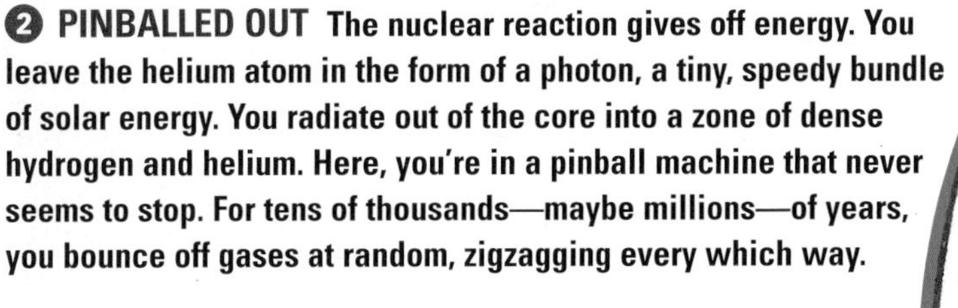

❷ PINBALLED OUT The nuclear reaction gives off energy. You leave the helium atom in the form of a photon, a tiny, speedy bundle of solar energy. You radiate out of the core into a zone of dense hydrogen and helium. Here, you're in a pinball machine that never seems to stop. For tens of thousands—maybe millions—of years, you bounce off gases at random, zigzagging every which way.

❸ SEEING THE LIGHT The collisions wear down your energy level. While stronger photons keep their high-energy gamma ray form, you weaken into X rays and then low-energy visible light. Finally, you reach a "cooler" part of the Sun—just 2,000,000 °K. There, gases swirl in superbumpy currents.

❹ RIDING THE RAPIDS You whoosh through the turbulent rapids to the surface—just 6,400 °K. From there, it's a swift trip through the Sun's atmosphere and into space. You and countless other energy particles travel in all directions away from the Sun.

❺ A SOFT LANDING About eight minutes after leaving the Sun, you enter the atmosphere of a tiny blue planet. High-energy photons such as gamma rays and ultraviolet light are blocked by the atmosphere. But you've cooled off. You can zip easily through the air and land on something soft. You are absorbed in the softness, as are many other solar photons. The cool thing starts to . . . melt.

Catch Some Rays

Here's a cool way to grab some Sun!

What you need:

- A sunny day
- A friend to help
- A piece of thin cardboard
- A piece of white paper
- A pen

What you do:

1. Using the tip of the pen, punch a hole in the thin cardboard.
2. Ask your friend to hold the cardboard in direct sunlight.
3. Hold the paper in the cardboard's shadow.
4. Look for a small, white circle. This is the upside-down image of the Sun!
5. Move the cardboard forward or back to focus the image.

When a cloud passes in front of the Sun, its shadow will cross the image in the opposite direction. (Never look directly at the Sun.)

Peak Performance

Mercury, Venus, Earth, and Mars are the Solar System's solid citizens—solid as rocks, as in rocky surfaces. That's why they're called the terrestrial or rocky planets. They're also grouped together as the "inner" planets. But in other ways, these four are very different, and one of the biggest differences among them is temperature.

Mercury is so close to the Sun that the part that faces the Sun is blazing hot. But because it has little atmosphere to keep heat in, at night—when the same area faces away from the Sun—it has extreme frigid temperatures. Venus is even hotter than Mercury, because it has a cloud cover that keeps the heat in—the so-called greenhouse effect. (Heat can get into a greenhouse, but not much escapes.) Mars is so far away from the Sun that it is a frozen wasteland. Only Earth has a temperature that's just right for water to exist as a liquid and an atmosphere that has enough oxygen to sustain life—at least life as we know it.

The differences among the terrestrial planets make for some strikingly different appearances.

▲ **Billions of years ago, asteroids and comets hit Mercury and the other planets during what we call the age of heavy bombardment. Earth was probably hit by them, too, but over time, air and water have scrubbed them off the face of our planet. Mercury has no air or water, so the craters could be there forever.**

Mars's Olympus Mons could easily cover the state of Wyoming. I t's almost three times taller (if you measure above sea level) than Mt. Everest, Earth's tallest mountain. So what's a massive mountain—and a volcano, too!—like Olympus Mons doing on Mars, a smaller planet than Earth?

Mars has a thick, frozen crust that can support a lot of mass without cracking. Earth has a thin crust. (Venus has a thinner one still.) But that's not the whole story. Because Mars is smaller than Earth, its gravity is weaker. Everything, including mighty Olympus Mons, weighs less on Mars. On Earth, Olympus Mons would weigh enough to crack through our planet's crust like an 18-wheeler on a newly frozen lake!

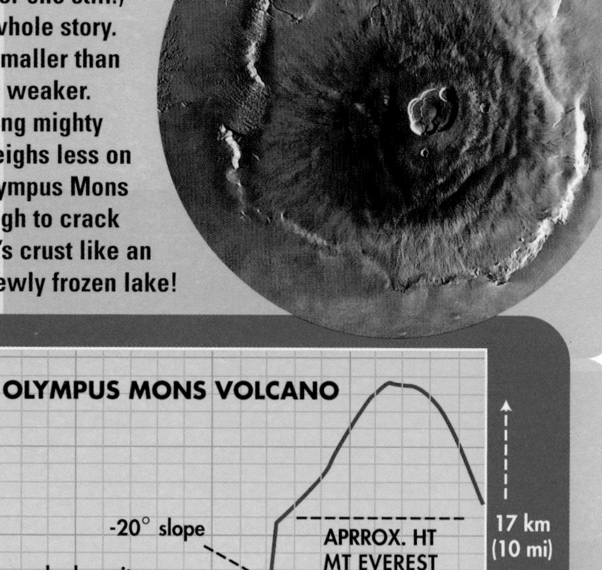

◀ Hot off the Venusian griddle are smooth, flat, and round "pancake domes." Heat softens Venus's crust into a thin skin. Magma easily breaks through the crust in countless volcanic "hot spots." Lava oozes as far as it can in all directions—and voilà! Venusian "breakfast" is served.

▼**Face Up:** Speaking of Mars, below is an actual photograph of a part of the Red Planet taken from the *Viking I* satellite in 1976. People have disagreed about exactly what it is. What do you think could have caused it?

Activities

Mount Chocolate All scientific models aren't exact, but some sure are sweet. One example: Chocolate candies such as Hershey Kisses™ respond a little like mountains on Venus and Mars. Heat makes them mushy and soft. Cold turns them brittle and hard. Don't believe it? Try the experiment below.

Freeze a few chocolates on a plate ("Mars"). Set others on a plate in direct sunlight ("Venus"). Leave a third batch at room temperature ("Earth"). Then put your mini-mountains to the test:

- Are they all the same "altitude" (height)?
- Have any changed shape?
- What other differences do you see?
- Tilt the plates. What happens?

Make Your Own
Make your own scrapbook of the Solar System. What should go in it? You decide!

GREETINGS FROM MARS

Go ahead—jump as high as you can. Now imagine leaping more than three times higher with ease. Your feet are as high as the roof of a car—then higher. Slowly, gently, you fall from this dizzying height and land on the frozen ground, unhurt.

With this giant leap, transport your mind to the surface of planet Mars. What would it be like to be standing there *right now*?

Superman on Super Mars?

Mars's weaker gravity makes you a superstar athlete—for a while. A week or so from now, your muscles will weaken without exercise. So you hop, leap, and skip over the surface of Mars to make them work hard and stay strong.

Soon, you're kicking up a layer of very fine orange dust. Oxygen in the ground literally rusted the dusty surface to a nearly uniform color.

The orange surface crunches and snaps loudly at each step. You carefully sidestep the sharp, scattered rocks, some partly hidden by dust. The rocks coat the plain, but in the distance, you see dunes, hills, and craggy crater walls on the impossibly close horizon. Mars is, after all, only half the diameter of Earth. It is also, you decide, a rusty, freeze-dried desert.

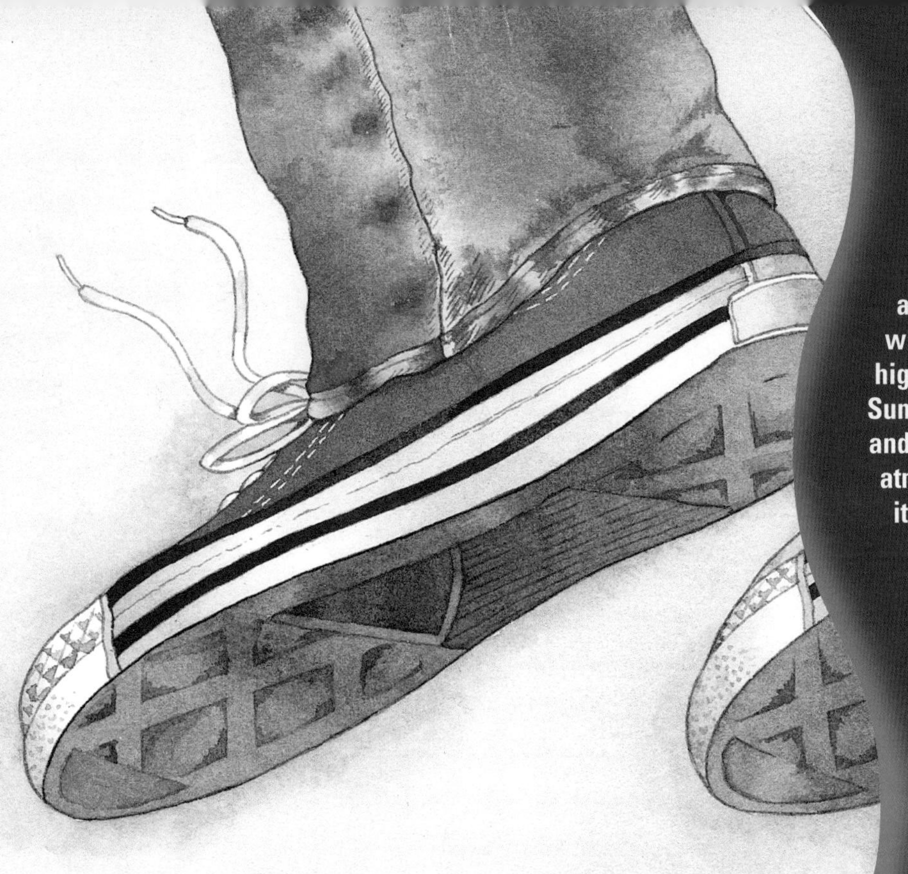

In the Pink

The sky is pale pink, beige, and light orange with a few wispy carbon dioxide clouds high overhead. You glance at the Sun and then look away. It's tiny and distant but very bright. The thin atmosphere can't shield you from its powerful radiation, so you avoid looking directly at it.

The atmosphere is almost all carbon dioxide, a greenhouse gas, but it is too sparse and spread out to hold in much heat. Because the gas is so thin, the air pressure is less than one one-hundredth of that on Earth. You hold out your arm to feel the wind. It's fast, a gale-force wind on Earth, but here the thin atmosphere makes it feel like barely a breeze.

One day is about the same as on Earth—24 hours from sunrise to sunrise. At sunset, the temperature will plunge to levels colder than those at the Earth's poles. For now, you enjoy the milder Siberian-like cold on Mars's warmest area—the equator. Still, you can feel the chill, even through your heated space-age jeans.

VENUS VACATION You're on Spring Break and your family decides to head to Venus. What would you do there? What would it be like? What would you have to bring? Soak in the Venusian facts in this book and other references. Then write a detailed description for your friends back home.

QUESTIONS & ANSWERS

JUPITER: He's a Gas!

Q: Hey, Big Guy. Any truth to the rumor that you're so big because you're full of gas?

A: What I want to know is, why does everyone point at me when they smell something funny? OK, it's true, but I'm not the only one. There are four of us called gas giants: me, Saturn, Uranus, and Neptune. I'm the biggest, but we're all HUGE. That's why another name for us is the jovian planets, after Jove, one of my nicknames. The four terrestrial guys are tiny compared to us. And they're all kind of clumped together near the Sun. We're WAY out here.

Q: Wait a minute. Four rocky and four gassy only makes eight planets. Our Solar System has nine. What about Pluto?

A: Yeah, right! What about him? Some scientists don't even think he's a planet at all. He's that small and out of it. My personal opinion? He's a moon of something else. But time will tell, time will tell. Anyway, let's keep the conversation where it belongs: on ME. Ask me exactly how big I am. Go on, ask me.

Q: OK. How big ARE you?

A: I'm 300 times heavier than Earth and more than twice as heavy as all the other planets added together. If Earth is a golf ball, I'm a basketball. How big is that?

Q: That's big, all right. How did you get that way?

A: How long do you have? It's a long story. Goes back more than four billion years, to when the Solar System was formed. Yep, we all came out of a cloud containing grains of dust and gas. The cloud spun and spun and the Sun started to take shape in the middle of it. And you know what that means.

Q: No, what?

A: Sun means heat. It was hot at the center of that activity—too hot for the stuff we're made of. Planets that formed closest to the Sun had metal cores because only metallic grains could take that heat.

Q: What kind of stuff are you made of?

A: Cold, gassy stuff like water vapor, ammonia, methane. Icy stuff. That's what we were formed around. Plus hydrogen, helium, carbon, nitrogen, and oxygen. Because of the cold, greater amounts of materials were able to condense in the outer planets. That's how we got so big: the bigger the planet, the more gravity it had to pull other stuff close and merge with it. Hydrogen and helium kept getting sucked in. And we got bigger.

Q: But . . . does that mean . . . you're not solid?

A: Mostly not. How did you know? Sorry about that. We gas guys don't have a solid surface. It's not like on Earth, where there's sky above and solid ground below. Our atmosphere just becomes denser with depth, as layer upon layer of it presses down. The boundary between it and my hydrogen "ocean" is kind of hazy. And then it's liquid all the way down to a tiny solid core. Tiny to me, that is. It's about the size of Earth!

Q: Really! What else is different about you?

A: We have rings. We have many moons. We're unbelievably cold. We're really far out.

Q: Didn't you say that already?

A: No, I said we are WAY out.

Q: Well, in any event, I think I get it now. But may I ask you one last question?

A: What?

Q: What IS that smell?

Activity

IT WEIGHS WHAT? Jupiter is only one-quarter as dense as Earth. That means if they were the same size, Jupiter would weigh only one-quarter as much as Earth. As we know, Jupiter is 300 times as heavy as Earth. If it were as dense, it would be 1,200 times as heavy!

To give you an idea of how the weight of a hollow object varies when filled with different substances, try this: Take two uninflated beach balls the same size. Weigh them—they should weigh the same. Cut a small slit in each one. Fill one with water and close the slit with sealing tape. Weigh the filled ball and subtract the original weight. This gives you the weight of the water.

Now fill the other ball in turn with other substances: sand, cellophane grass, foam peanuts. Seal the filled ball each time with strong but peelable tape. Weigh it, and remember to subtract the weight of the ball. How many times heavier or lighter is the water than the other materials?

Note that this is not a true test for the density of the materials used, because in the case of the foam peanuts, for example, there is a lot of air between pieces. But it gives you an idea of how objects, the same size can vary so much in weight.

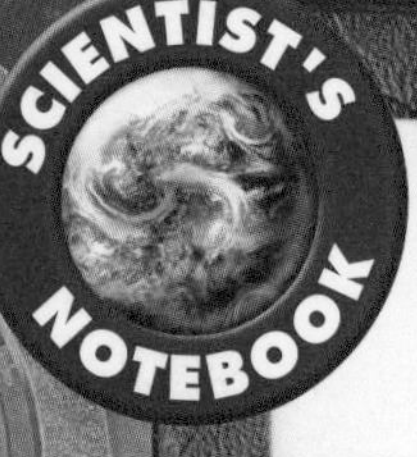

Welcome to the Family!

For thousands of years until 1781, people thought there were just six planets in our Solar System: Mercury, Venus, Earth, Mars, Jupiter, and Saturn. That's because there was no way of seeing planets seven, eight, and nine. Let's have a look at what could have been some diary entries from the times of their discovery.

Bath, England, Spring 1781

I Didn't Planet That Way

Observer: William Herschel

Instrument: 6.2-inch (16-cm) homemade telescope

Location: Near the stars forming the left foot of Castor—one of the "twins" that form the constellation Gemini

Observation: March 13: At about 10 P.M., spotted curious smudged star or perhaps a comet. Records show that 20 past observers, since 1690, consider it a star.

March 17: Not a star! It's a disk, more like a planet, but I think it's a comet because it changed place.

April 15: French astronomer Charles Messier says it's not like any comet he's seen. Could it be a planet? But there are only six planets.

April 20: Russian astronomer Anders Lexell calculated the orbit. It's a planet!

April 26: I announced discovery to the Royal Society. I shall call it the Georgian Planet, after King George III.

Six decades later: The name Uranus—after the first ruler of the universe in Greek mythology—won out over Georgian Planet, Planet Herschel, and Planet Neptune.

Uranus

Neptune

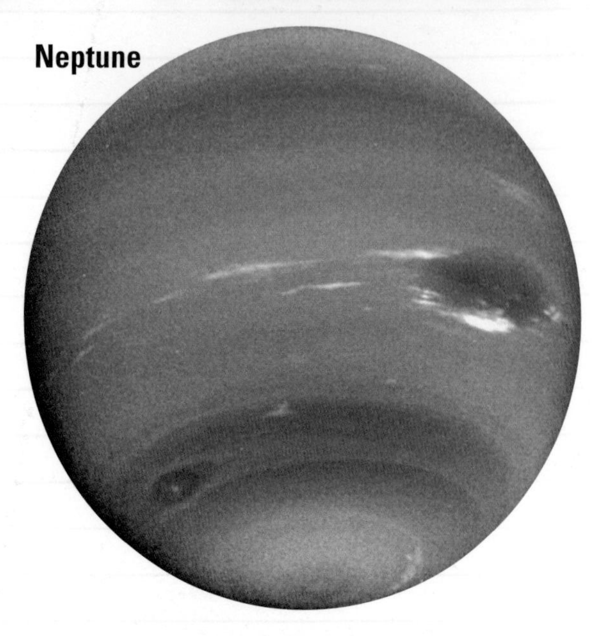

Cambridge, England, 1840s

It's Gotta Be There

Observer: John Couch Adams

Instrument: Paper and pencil

Observation: June 1841: The new planet Uranus follows an irregular path. Could another planet be pulling it off its course? Will calculate Uranus's orbit.

October 1843: Calculations complete. There must exist an eighth planet beyond Uranus! Will refine figures to predict exact location.

November 1845: Figures rejected by Royal Astronomer. No one will point a powerful telescope at the unseen planet's predicted spot.

Pluto

Charon
(moon of Pluto)

France, 1846

Observer: Urbain Leverrier
Instrument: Paper and pencil
Location: The constellation Aquarius
Observation: June: Calculated position of suspected eighth planet. To affect the orbit of Uranus, the planet must be as far from Uranus as Uranus is from the Sun.

August 31: Calculated the size of the missing planet. Must find someone with a powerful telescope.

September 18: Finally! Johann Galle has agreed to look for my planet.

Germany, 1846

Observer: Johann Galle
Instrument: 9-inch (23-cm) telescope
Location: The constellation Aquarius
Observation: September 23: I found it! Pointed telescope at Leverrier's coordinates and checked stars against a star map. Spotted a faint light where no star should be.

Fast Forward: The new planet was named Neptune after the Roman god of the sea. Galileo probably spotted Neptune in 1613, when its position as seen from Earth was relatively near that of Jupiter, but he didn't know it was a planet. In 1989, *Voyager 2* revealed that Neptune is a blue icy planet with fierce winds.

Flagstaff, Arizona, U.S., 1930

You Missed a Spot

Observer: Clyde Tombaugh
Instrument: 13-inch (33-cm) telescope, blink comparator (special kind of microscope)
Location: The constellation Gemini
Observation: January 23: Joined Lowell Observatory, where we are looking for "Planet X," a ninth planet. The telescope here sure is swell! Beats my homemade reflector made from farm machinery and car parts. Took one-hour photo of a section of Gemini.

January 29: Took another photo of the same section. Boss looked at photos but didn't see anything interesting in all those smudges of light.

Clyde Tombaugh

February 18: At four o'clock, compared two photos taken last month. Spotted a very dim light that moved just an eighth of an inch out of place. Hard to contain my excitement, but I've got to be sure.

March 13: It's a planet!

Fast Forward: The planet was named Pluto after the Roman god of the underworld. It is too small of a planet to affect the orbits of Neptune and Uranus. A moon, Charon, was discovered in 1978.

Activity

DEAR DIARY Imagine what it would be like to discover a planet in the Solar System. What kind of media coverage do you think there would be? Write a diary account of the first year after your discovery.

MEGA MOONS

Moons may not get the same press as planets, but many are too massive to be ignored. In fact, seven moons are bigger than Pluto, and two of those are even bigger than Mercury.

You remember that Galileo spotted the first four of Jupiter's moons with his telescope. As people developed more powerful instruments for space study, more of the planets' moons came into view.

Only five of Uranus's moons were known before 1986, when *Voyager 2* spotted ten more. In late 1997, two more moons were seen by powerful telescopes, and an eighteenth in 1999 was the result of a reexamination of a 1986 *Voyager* image. The first fifteen moons are named for characters in plays by Shakespeare, but not always appropriately. Miranda, named after the young beauty in *The Tempest*, has a bizarrely scarred surface.

HERE ARE ALL THE KNOWN MOONS OF THE SOLAR SYSTEM:

EARTH

1 Moon

MARS

2 Phobos, Deimos

JUPITER

16 Metis, Adrastea, Amalthea, Thebe, Io, Europa, Ganymede, Callisto, Leda, Himalia, Lysithea, Elara, Ananke, Carme, Parsiphae, Sinope

SATURN

18 Pan, Atlas, Prometheus, Pandora, Epimetheus, Janus, Mimas, Enceladus, Tethys, Telesto, Calypso, Dione, Helene, Rhea, Titan, Hyperion, Iapetus, Phoebe

URANUS

18 Cordelia, Ophelia, Bianca, Cressida, Desdemona, Juliet, Portia, Rosalind, Belinda, Puck, Miranda, Ariel, Umbriel, Titania, Oberon, plus 3 unnamed

NEPTUNE

8 Naiad, Thalassa, Despina, Galatea, Larissa, Proteus, Triton, Nereid

PLUTO

1 Charon

GALLERY OF MOONS

This monster moon, Ganymede, is the biggest in the Solar System.

Welcome, craters! Callisto's heavily cratered surface comes from being slammed by meteoroids.

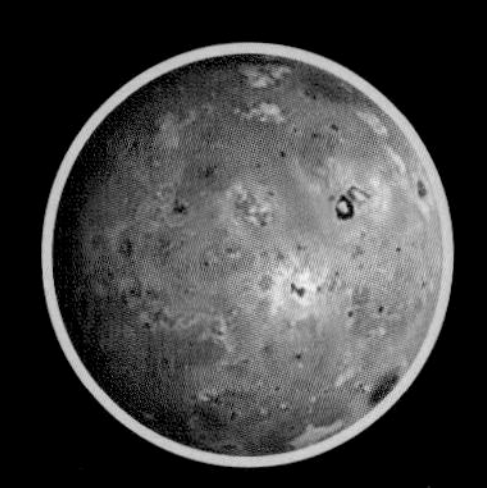

Resembling an extra large pizza with cheese, Io bubbles with active volcanoes and fresh lava flows.

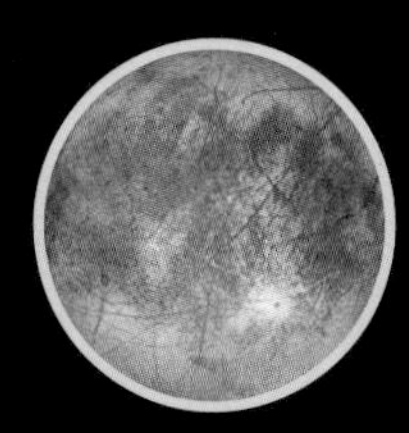

Can you see the sea? Some scientists think there may be a liquid ocean beneath the icy crust of Europa.

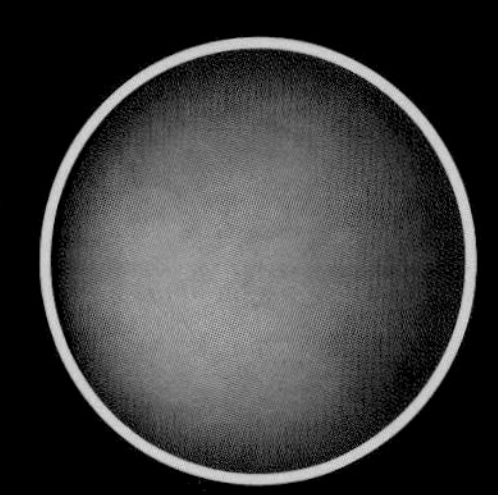

Larger than Mercury, Saturn's Titan is the only moon with an atmosphere.

Look out below! Someday, Triton will crash into its planet, Neptune.

Earth's Moon is considered huge for the size of its planet.

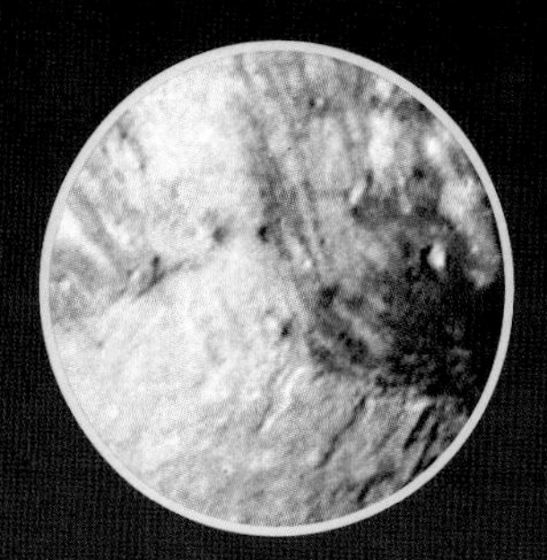

Miranda is the heavily scarred moon of Uranus.

Activity

MOON MAGIC Choose a moon (ours included) that really interests you, and write a log of a trip you would take there if you could. Include information such as: How you would get there, and what it would be like? (Use lots of details.) What kind of special equipment would you need to survive? If the moon you pick is not our own, research its name too.

Space Hunt

***G**oliath* **is an imaginary space probe, but the information we're receiving from it models what scientists actually study over millions of miles from space.**

Goliath, where are you?

Here at the NASA Jet Propulsion Laboratory in Southern California, we're tracking our imaginary space probe, *Goliath*. It's an incredible machine that zips around the Solar System, sending back all kinds of telemetry to Earth. What the heck is telemetry? It's data zapped from one place (such as a probe in space) to another (such as a computer on Earth). Telemetry includes temperature readings, speed of the probe, digital images, and other information.

Unfortunately, there isn't a simple headline that tells us which planet or moon *Goliath* is exploring. Can you figure it out?

To solve the mystery, first study each piece of telemetry data. Then refer to each corresponding clue in the Detective's Notebook. Remember: Think like a detective, and use the information to rule out wrong answers right away.

(Answer is on page 32.)

TELEMETRY

- 1:20 P.M. Pacific Standard Time (PST): *Goliath* radio transmission begins from unknown location in the Solar System.
- 2:32 P.M. PST: Radio signals begin to arrive at the JPL
- Current speed of Goliath: 40,325 mph. (64,883 kph)
- Location: Approaching object that has a diameter of 3,199 miles (5,150 km)
- Density of object: 1.9 grams per cubic centimeter
- Gas chromatograph detects atmosphere of 90 to 98 percent nitrogen, 2 to 10 percent methane, traces of organic (carbon-based) molecules
- Amount of sunlight that reaches surface: 10 percent
- Surface temperature: 94 °K (–290°F or –179°C)

Clues

Detective's Notes

First make a list of the other eight planets besides Earth.

- Assume the planets are lined up in one direction from the Sun.
- To figure out approximately how far away Goliath is, figure out the number of seconds from the time radio transmission starts till the time the first signals arrive on Earth. (Hint: speed of radio signals is 186,282 miles per second (299,728 kps).)
- What does the speed of Goliath have to do with solving the mystery? Could it be a bogus clue?
- Which planets have this diameter? If none is a match, try a moon.
- This density suggests object is half-rock, half-ice.
- Which planets or moons have an atmosphere? Which ones have an atmosphere that includes nitrogen and methane?
- Do you need the clue about the amount of sunlight? If you think so, where would you find this information?
- At 94° Kelvin, methane can be in gas, liquid, or solid form. Methane clouds, methane rain, methane lakes. Methane glaciers?

ARE WE DOOMED?

Will the Earth be destroyed?

The short answer is yep. Odds are that a huge asteroid or speeding comet will slam into Earth and destroy all life and perhaps the planet too. Even if that doesn't happen, our Sun will swell into a red giant star about five billion years from now. It will evaporate the oceans and end all life on Earth. Yikes!

But the long answer is more complicated—and more hopeful. Most asteroids burn up in the atmosphere without reaching the ground. But there are notable exceptions. On June 30, 1908, a comet or an asteroid exploded in the air over Siberia, Russia. It flattened a forest for miles in all directions, leaving a Paul Bunyan version of a "crop circle." Plenty of people witnessed the fireball. About 50,000 years ago, an asteroid about the size of a football field created Meteor Crater in Arizona. The mile-wide (1.6-km) hole dwarfs tourists who flock to see it. On average, objects as big as this hit Earth roughly once in several thousands of years.

These two impacts are nothing compared to the last really big impact. To kill off species, an asteroid needs to be at least half a mile (about a kilometer) wide. These medium-sized smashers tend to visit Earth about every 300,000 years or so. They can create monster tsunamis or flatten plenty of miles of real estate.

Really big asteroids, about 6 miles (10 km) in diameter, strike roughly once every 100 million years or so, causing mass extinctions. They kick up so much dust and debris in the atmosphere that little sunlight can get through. Plants die and, soon, some animals that eat the plants die. The space rock that many scientists think helped kill off the dinosaurs 65 million years ago was probably about 10 miles (16 km) wide.

When will one of these killers hit? No one knows. In a minute or not for millions of years. Meteorites don't fall on schedule. But they do fall, as the map shows. Even though wind, rain, and volcanoes have worn away traces of most of Earth's impact craters, scientists have pinpointed where they've landed. Usually, they find rocks that only a high-speed impact can create. For example, tektites form when molten rock is tossed into orbit and then solidifies into glassy rock on the way down. Also, some asteroids contain an element called iridium that's rare on Earth.

The first challenge to saving Earth from a meteorite is to see it coming. Astronomers are scouring the skies for big and medium asteroids and comets that cross Earth's orbital path. Really big near-Earth asteroids are rare, but hundreds of medium-sized ones exist. The next step is to avoid or deal with the impact.

This map shows where 140 impact craters have been identified on Earth. They range in size from about one-half mile (about 1 km) to over 22 miles (35 km) across. Some go back 2 billion years. Do you think you are safe living near or far from existing disasters?

Activity

A Fine Mess A scientist has discovered that a killer asteroid will hit Earth in a matter of weeks. But no one knows exactly where. What would you do if you were in charge of preparing the world for the hit? Create a three-part plan to divert the space rock, lessen its impact, and save as many lives as possible in the event of a direct hit. Compare your plan with the plans of other groups. Set up a debate to argue their merits.

HERE COMES

Halley's Comet zooms by Earth regularly, and so it has a chance—a tiny one—of colliding with our planet. During a flyby, the famous chunk of ice is easy to spot, even without a telescope. For example, there are reports from 240 B.C., when ancient Chinese stargazers saw it without any problem.

Here's a time line that shows some of the world-changing events that happened between Halley's visits. As you will see, Halley doesn't appear at exactly even intervals. By computing the average time between visits, what year would you expect the comet to turn up next?

1301

1325 Aztecs build Tenochtitlán (now Mexico City).

1338 Hundred Years War starts between England and France.

1341 Black Death begins in Asia, reaches Europe in 1348.

1368 Mongols expelled from China, Ming Dynasty begins.

1378

1400 Geoffrey Chaucer, author of *The Canterbury Tales,* dies, his great work unfinished.

1405 Chinese sailors explore the Indian Ocean.

1431 Joan of Arc burned at the stake.

1455 Gutenburg invents printing press.

1456

1492 Columbus sets sail to find a route to Asia through the Pacific.

1502 First slaves arrive in the Americas.

1503 Leonardo da Vinci paints the Mona Lisa.

1512 Michelangelo completes the painting of the Sistine Chapel ceiling.

1521 Martin Luther excommunicated by Roman Catholic Church.

1531

1542 Copernicus formulates theory of a Sun-centered Solar System.

1545 Silver is found in the Andes Mountains of South America.

1558 Queen Elizabeth I of England begins forty-four-year reign.

1607 English settlers arrive at Jamestown, Virginia.

1607

1609 In Germany, first regularly printed newspaper begins publication.

1610 Galileo Galilei, through his telescope, first to see some of Jupiter's moons.

1618 Thirty Years War begins.

1620 Pilgrims arrive in Plymouth, Massachusetts, aboard the *Mayflower.*

1646 Edmund Halley, for whom comet is named, born in England.

Author Mark Twain was born in 1835 when Halley's Comet was shooting across the sky. He said, "It will be the greatest disappointment of my life if I don't go out with Halley's Comet." On April 21, 1910, the day after the comet returned, Twain died.

HALLEY

1682	1759	1835	1910	1986
1692 Salem witchcraft trials. **1717** Scottish and Irish immigration to the American colonies begins. **1742** Edmund Halley dies. **1752** Benjamin Franklin, flying kite in thunderstorm, proves lightning is electricity. **1754** French and Indian War begins.	**1775** American Revolution begins. **1781** Uranus discovered. **1789** George Washington elected first U.S. president; French Revolution begins. **1800** Alessandro Volta invents the battery.	**1842** Anesthesia first used in surgery. **1846** Neptune discovered. **1861** Civil War begins. **1867** Dynamite invented by Alfred Nobel. **1876** Alexander Graham Bell makes the first telephone transmission. **1903** Orville and Wilbur Wright build an airplane that flies for twelve seconds. **1908** Henry Ford rolls out his first automobile.	**1914** World War I begins. **1920** Women gain right to vote in U.S. elections. **1928** General Electric makes the first television broadcast; penicillin is discovered. **1930** Pluto discovered. **1939** World War II begins. **1945** Atomic bombs dropped on Hiroshima and Nagasaki, Japan. **1969** Neil Armstrong walks on the Moon. **1973** Vietnam War ends.	**1986** First official U.S. observation of Martin Luther King Jr. Day; space shuttle *Challenger* explodes. **1989** *Voyager 2* passes Neptune and leaves the Solar System. **1990** Hubble Space Telescope launched. **1991** Soviet Union breaks up. **1995** U.S. space shuttle *Atlantis* docks with orbiting Russian space station Mir. **1997** First cloned mammal is sheep. **2003** *Columbia* space shuttle breaks apart during re-entry.

Activity

Hello Again, Halley! Halley's Comet is back. How old will you be? What will your life be like? Make a mock-up of the front page of a newspaper that might be published on the day Halley's Comet returns.

Carolyn Shoemaker, Meteor Master

Flagstaff, AZ, 1983–Present

What if there were a doomsday meteor hurtling toward Earth right now? Wouldn't it be great if there were a hero who could spot it in time to stop it? That hero could very easily be Carolyn Shoemaker. Since 1983, she's discovered more than 800 asteroids and 32 comets, some of which cross Earth's orbit.

Ms. Shoemaker is already a hero in the eyes of world astronomers. Working with her husband, Gene, and David Levy, she discovered Comet Shoemaker-Levy 9—or what was left of it.

Jupiter's gravity had grabbed the comet and forced it to orbit the gas giant instead of the Sun. The powerful gravity ripped the comet into 21 pieces, all of which slammed into Jupiter in 1994. The spectacular crash gave astronomers an exciting glimpse at the chemistry of Jupiter's mysterious atmosphere.

What do you say when you're the first human to set eyes on a comet? In the words of Carolyn Shoemaker, "Yay-y-y-y!"

Below is part of an interview with one of the world's best meteor spotters.

There is no way I can claim full credit for these discoveries. It's very much a team effort.

It was never my intent to go into science. I had intended to be a teacher . . . [but my husband] Gene suggested a project that needed a little help. It had to do with the discovery of asteroids and near-Earth objects. I enjoyed finding asteroids on the plates taken with the Palomar 18-inch Schmidt [telescope]. It was like looking at the sky through a window that really put me out there with the stars.

We look at our Palomar films two at a time, side by side on the stereomicroscope. Typically, they are taken 45 minutes to an hour apart. One eye looks at one film and the other eye looks at the other film, and the brain happily melds the two into one. This has a wonderful, magical effect: Asteroids and comets appear to float above the flat surface of the stars.

When I found my first near-Earth object in 1983, I was really excited. [It looked like a] little pair of hyphens. Then I was hooked.

On our films were galaxies and all sorts of beasts in the zoo out there.

Then I found my first comet. It was just a pure thrill. I didn't know much about comets and I hadn't been hooked on them, but it only took one. I knew I just had to find others.

People wonder if after [so many] comets you get the same thrill. All I can say is that every time I find a comet, I want to dance. I usually call up to the observers at the telescope and yell, "Yay-y-y-y!"

Comet Shoemaker-Levy 9 crashing into Jupiter.

CAREERS

Are there stars in your skies?

Would you like to become a star watcher? Not the kind who hangs around the set of a Hollywood movie, but one who looks at the heavens. This is part of what astronomers do. They study whole galaxies, as well as planets, moons, and other residents of the Solar System. Their research can cover such topics as the origins of the universe, how galaxies form and change, and the structure of stars and planets—things too far away to see without sophisticated scientific instruments. And they sometime study the physics of particles too small to see, even with the help of the best technology.

If working as an astronomer sounds interesting, you may want to check out what you need to do to become one. You need at least an undergraduate college degree, and most likely a doctor's degree. A thorough knowledge of astronomy, math, and physics is essential, and you have to keep up to date with new information and discoveries. You have to love and be good at research, because you're going to have to come up with good ideas for new projects. You'll need to be self-motivated, detail-oriented, and accurate, and have good organizational, technical, computer, and problem-solving skills.

Astronomers study the surface of rocky planets, and so do geologists. The general requirements for being

a geologist are similar to those for an astronomer—it's just that the scientific focus is different. Geologists studying rocks that spacecraft bring back from Mars or fragments of space rocks that fall to Earth help us understand how the Solar System may have formed billions of years ago.

In yet another space-age job, as an aerospace engineer, you may help design a space probe that gathers data and samples in places no human can venture. An aerospace engineer has to be excellent at solving problems that the harsh conditions of the Solar System create. Aerospace engineers often have undergraduate degrees in math or physics, or electrical or mechanical engineering. They may have a degree as high as a doctorate in aeronautics or rocket science.

But even if you have no desire to make a career in space study, you can still enjoy yourself aiming a telescope at the heavens in your backyard or in an observatory. After all, we are made of stardust!

FUN & FANTASTIC SOLAR SYSTEM Lite

The Name Game

Though English is generally recognized as the international language for professional astronomy, other countries call the Sun by another name in their everyday language. How many can you pronounce? Do any astronomical sounds come from these names?

Spanish and Latin:	**Sol**
French:	**Soleil**
Italian:	**Sole**
German:	**Sonne**
Greek:	**Helios**
Japanese:	**Taiyou**
Korean:	**Taeyang**
Hungarian:	**Nap**

Changing Lineup

Every 238 years, Pluto's shaky orbit path brings it closer to the Sun than the orbit path of Neptune. Thus, depending on how far along it is in its orbit, Pluto can be either the eight or ninth planet from the Sun.

TOUCHDOWN!

A space shuttle landing takes more than thirty minutes. The return to Earth begins when the astronauts slow the craft and ease it into Earth's atmosphere by using small rockets in the craft's nose and tail. After re-entry, they steer it like an airplane, using rudders and flaps. Here it slows down from 1,000 to 100 meters (about 3,300 to 330 feet) per second and withstands temperatures of up to 1,800°F (about 1,000°C).

Ideally, each shuttle mission ends with a landing at Florida's Kennedy Space Center. However, when weather conditions are poor there, mission control directs the shuttle to land on a dry, desert lake bed at Edwards Air Force Base in Southern California. From there, the shuttle rides back to Kennedy atop a Boeing 747. After being checked, it will be ready for another lift-off..

Star Tours

Q: Are there really shooting stars?

A: Not exactly, but meteoroids traveling in outer space heat up and glow as they fall through Earth's atmosphere. They're called "falling stars" as they fall to Earth.

Weighty Matters

- **The entire atmosphere weighs 5,700,000,000,000,000 tons—that's 5,700 trillion tons!**
- **The largest meteorite ever found on Earth fell in Namibia, Africa. It originally weighed 100 tons (91 tonnes).**

Distorted Views

Through a telescope, Saturn looks squished. That's because it's mostly gas and liquid, and it becomes slightly compressed as it spins very quickly.

Notable Notes

- Though Uranus was discovered in 1781, the only space expedition to it was *Voyager 2*, which flew by in 1986.
- Mercury whizzes around the Sun at 30 miles (about 50 km) per second.
- If you are twelve years old on Earth, you'd be only one year old on Jupiter.
- Venus is not only the closest planet to Earth, but also the brightest as seen from Earth. That's why you can sometimes see it in daylight.
- Orbitally speaking, the average Earth year is actually 365 days, 5 hours, 48 minutes, and 45.51 seconds. The average Earth day is 23 hours, 56 minutes, and 4.09 seconds.
- Saturn's ring particles vary in size from that of a grain of sugar to the size of a house.
- When a comet approaches the Sun, its tail is following; when it moves away from the Sun, its tail is leading. That's because of the pull of the Sun's gravity.
- If you're on the equator, you can view all of the constellations over the course of a year. If you are on the North or South Pole, you will be able to see only one-half.
- The Solar System orbits the galaxy about once every 250 million years. Thus, it has gone around the galaxy only 15 to 20 times.

See for Yourself: Frequently Asked Questions

1. Why do stars twinkle?
2. How many stars can you see on a clear night?
3. Which planets can you see without a telescope?
4. Why does the Moon shine?

Answers:
1. Because starlight comes down to Earth through different layers of air.
2. Almost 3,000 with just your naked eye.
3. Mercury, Venus, Mars, Jupiter, and Saturn.
4. The Sun's light reflects off the Moon's surface.

War of the Worlds

On October 30, 1938, mass hysteria hit the United States when Martians invaded New Jersey . . . or at least that's what people believed. An hour-long radio broadcast by actor Orson Welles and his *Mercury Theater on the Air* brought an 1894 book by H. G. Wells, called *War of the Worlds*, to life.

In a mock news broadcast that broke into the station's regular music hour, the audience was told of an invasion by "monsters" attempting to wipe out civilization, beginning with New York and New Jersey. Though an announcement stating that it was a fictional play ran four times, many listeners tuned in for only a short time, heard the message and reacted immediately. Police switchboards lit up across the country, and many people on the East Coast ran into the streets or called loved ones in panic, thinking they would not have long to live.

MOON MAN **Have you ever heard of the man in the Moon? For generations, some Americans have claimed to see a human face in the crater-scarred Moon. Other cultures have passed on their own tales about the figure. Native Americans tell a story of seeing a frog trying to protect the Moon from a bear that wants to swallow it. Scandinavian children are taught a folktale about Jack-and-Jill–type characters, Hjukl and Bill, who are holding a pail of water and tumbling down a hill as they are trying to run from their father.**

Final Project: Design Your Own Solar System

If you could live in a different solar system, what would you want it to be like? It's your world—and your turn.

In the 1990s, astronomers discovered twenty planets around stars other than our Sun. Each planet is part of a solar system that looks almost nothing like ours.

One system has at least three planets, all bigger than Jupiter! The two outer giants swing close to their young, Sun-like star and then zoom way out. Their mighty gravity would wipe out any puny Earthlike planets in the system.

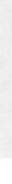

Imagine just one of the 200 billion stars in the Milky Way. Then use your knowledge of space science to design a workable solar system. Draw a map and describe the star, planets, and moons in captions.

Some Thrilling Possibilities

- What type of star is at the center of your system?
- How many planets and moons does your solar system have? How big are they? Are they gas giants or terrestrial?
- How far apart are the planets spaced? Are their orbits almost circular or very elliptical?
- Could any planets support life? How?

ANSWERS

Solve-It-Yourself Mystery, page 22: The answer is Titan, one of Saturn's moons, and here's how to figure out the answer: The quickest way is not to go piece of data by piece of data (and clue by clue) in the order given but to see which ones could be the real problem solvers. In this case, the diameter is the first key. If you look through our almanac, you discover that none of the planets has a diameter of 3,199 miles (5,150 km). So *Goliath* must be approaching a moon. Check out the information on our Mega Moons pages. Titan is listed as the only moon with an atmosphere. You've solved the mystery without doing any calculation! A harder way to solve the mystery would be to calculate (by dividing the speed of radio signals into the difference in time between the start of radio transmission and the arrival of signals, to give distance from Earth, then adding the distance from the Earth to the Sun) that *Goliath* is in the vicinity of Saturn. Since Saturn is obviously much, much bigger than 3,199 miles in diameter, the object must be one of its moons. Again, Titan is the only moon with an atmosphere, so your mystery is solved once again. The rest of the data would be necessary only if more than one moon fit the earlier data.

Here Comes Halley, page 26:
About 2061. The period between visits is usually 76 years, but ranges from 74 to 79 years. The reason? The comet sometimes passes close to Jupiter and Saturn, which have strong enough gravity to slightly change its course and speed.